Earth

Andrew Charman

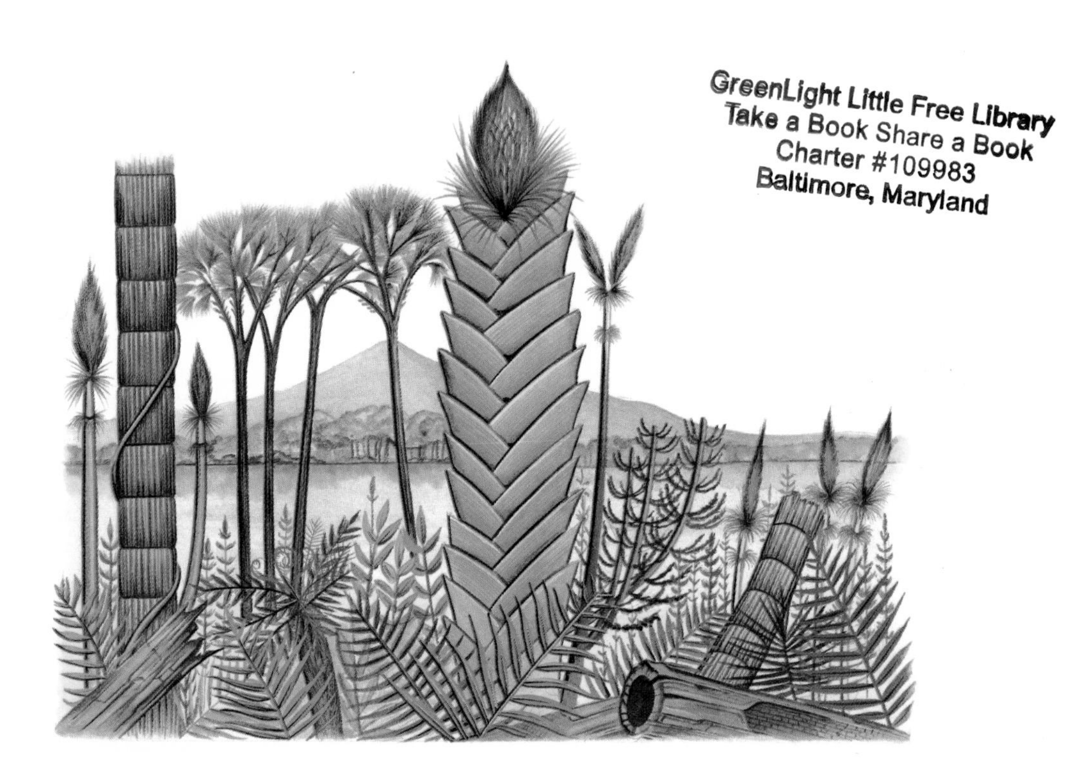

RAINTREE
STECK-VAUGHN
PUBLISHERS
The Steck-Vaughn Company

Austin, Texas

Series Editor: Pippa Pollard
Editor: Claire Llewellyn
Design: Shaun Barlow
Project Manager and Electronic Production: Julie Klaus
Artwork: Ray Grinaway
Cover artwork: Ray Grinaway
Picture Research: Ambreen Husain
Educational Advisor: Joy Richardson

Library of Congress
Cataloging-in-Publication Data
Charman, Andrew.
Earth / Andrew Charman.
p. cm. — (First starts)
Includes index.
Summary: Describes the planet Earth, its plates, rocks, minerals, and fossils. Includes mining, the breaking up of rocks, and all uses and features of soil.
Hardcover ISBN 0-8114-5510-6
Softcover ISBN 0-8114-4923-8
1. Earth — Juvenile literature. 2. Geology— Juvenile literature. 3. Soil physics — Juvenile literature. [1. Earth. 2. Geology.] I. Title. II. Series.
QB631.4.C48 1994
551—dc20 93-31791
CIP
AC

Printed and bound in the United States by Lake Book, Melrose Park, IL

2 3 4 5 6 7 8 9 0 LB 98 97 96 95 94

Contents

Planet Earth

The planet on which we live is a huge ball-shaped lump of rock spinning through space. The surface of this ball is under attack — from wind, rain, and heat. They break up the hard rock into smaller pieces. This is what makes soil. Soil is very important to life on Earth. Most plants cannot grow without it, and animals cannot live without plants.

▽ There are many different kinds of landscapes in the world. Underneath the soil lies hard rock.

Under the Surface

▷ Erupting volcanoes are holes in the Earth's surface. Molten rock from below pours out of them.

Deep down inside the Earth, it is so hot that the rocks have melted. The land on which we live forms part of huge pieces of rock, called **plates**, which float on this molten rock. Although we rarely feel it, the plates move very slowly and bump into each other. Over millions of years, this has pushed up mountains and made the oceans change shape.

▽ The Earth's surface is divided into plates. When two plates grind past each other, then earthquakes happen.

▽ The plates of rock also bump into each other. This pushes up mountains.

The Rocks Around Us

There are three main kinds of rocks on Earth. Igneous rocks are formed from cooled **lava**. Sedimentary rocks are made of layers of sand, mud, and plant and animal remains that have been squeezed together and become hard. Metamorphic rocks are formed when the other two kinds are changed by heat, **pressure**, or **chemicals**.

▷ Sugar Loaf mountain is made mainly of granite, a hard igneous rock.

▽ Sandstone is a soft, sedimentary rock. Sandstone cliffs will gradually crumble away.

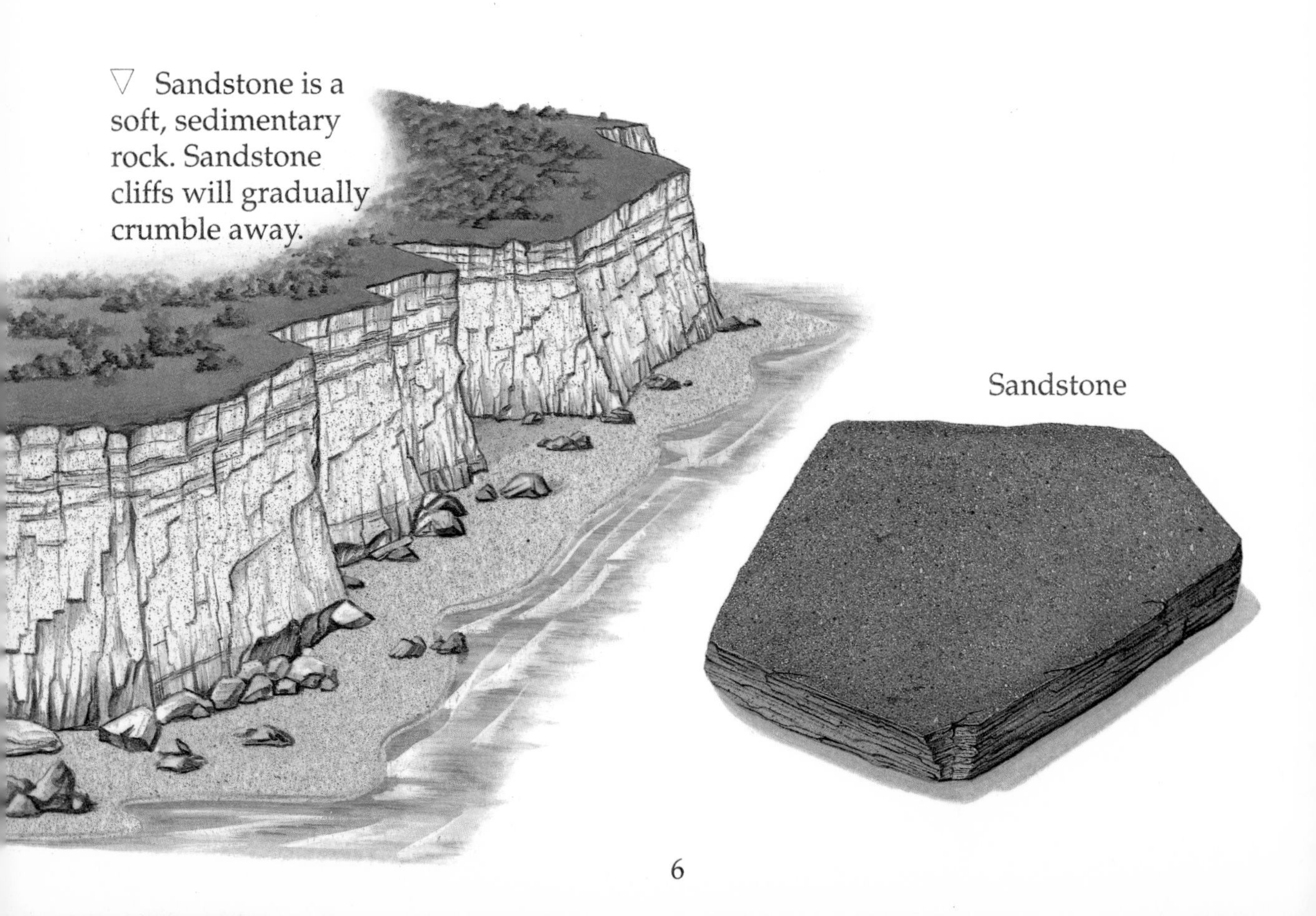

▷ Marble is a metamorphic rock and can be polished.

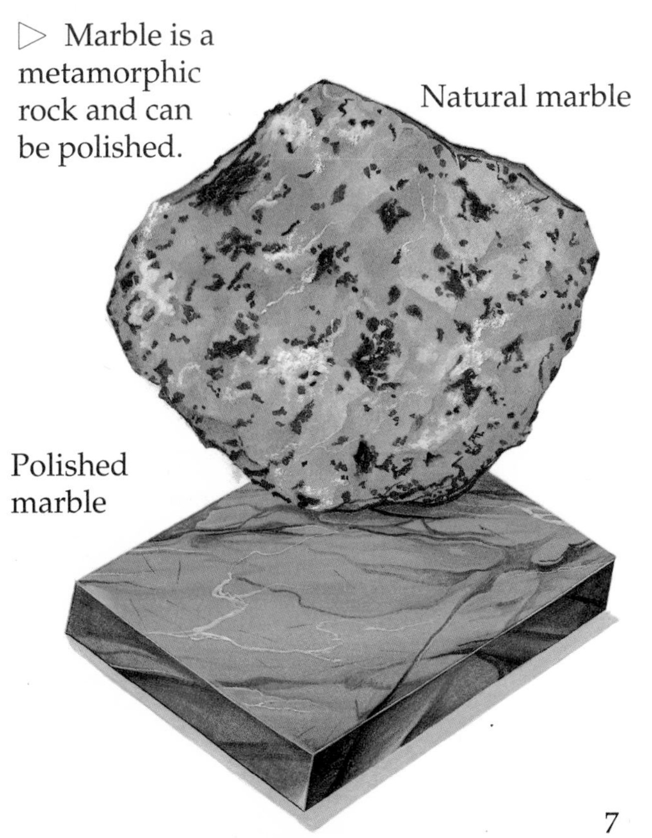

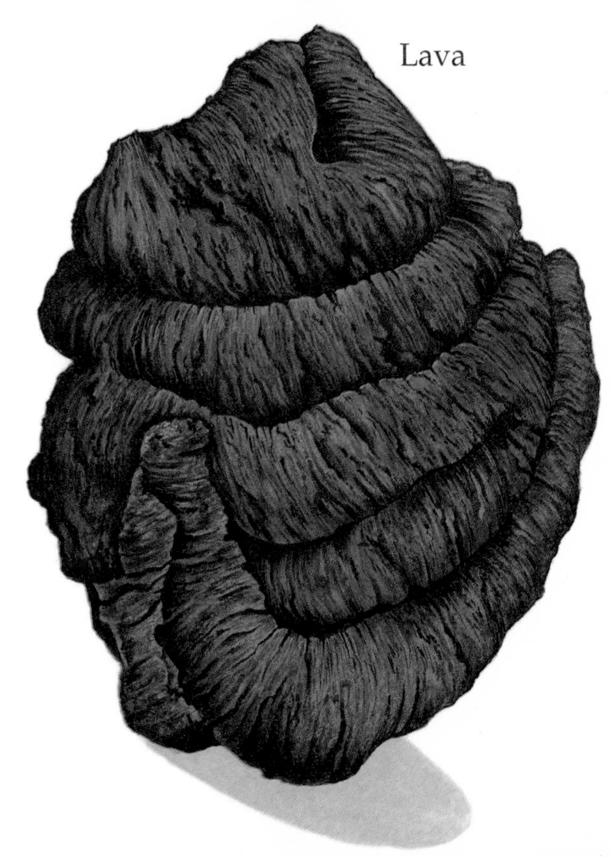

Minerals in the Earth

Rocks are made of substances called **minerals**. There are thousands of different kinds of minerals. Some rocks are made of only one kind of mineral, others are made of several. Some minerals form gemstones. Diamonds, rubies, sapphires, and opals are all examples of gemstones. They are treasured because they are beautiful and rare.

▽ Opals form in cracks in the rocks. Here, opals are being mined for use in jewelry.

▷ Jade and turquoise are gemstones. They are valued for their color.

◁▷ Some minerals, called ores, contain metals. The ores are crushed so the metal can be taken out.

Fossils in the Rocks

Some rocks contain fossils. These are traces of animals and plants that lived millions of years ago. After an animal died, layers of rock formed on top of its body. Minerals seeped into the rock and replaced its bones or shell. Plants have been fossilized in the same way. The **fuel** called coal is the fossilized remains of plants that grew long ago.

▷ The fossilized bones of a dinosaur are sometimes found in rocks.

▽ This is the fossil of an ammonite. These animals lived in the ocean millions of years ago.

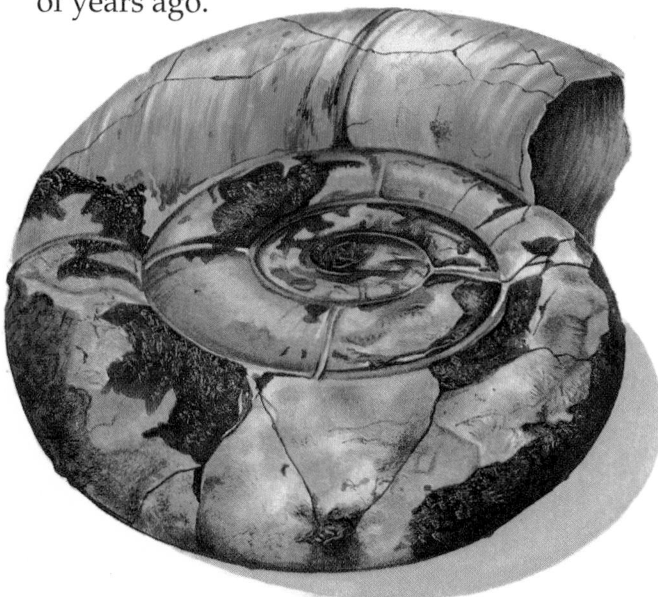

▽ Coal is the pressed remains of swampy forests that grew millions of years ago.

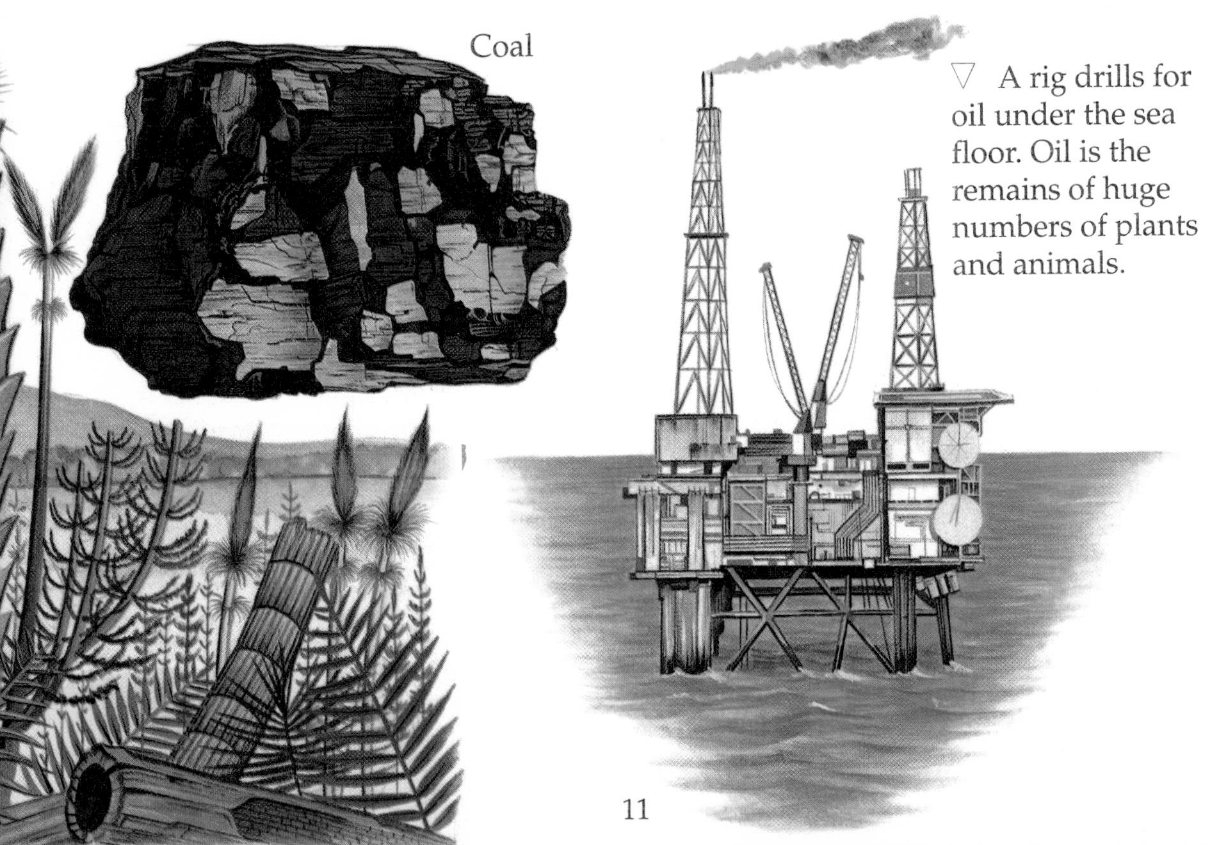

▽ A rig drills for oil under the sea floor. Oil is the remains of huge numbers of plants and animals.

Mining the Earth

▷ This machine is being used to dig tunnels in a coal mine. Underground mining is dangerous work.

Mining is taking materials from the ground. Metal ores, gemstones, and coal are all mined. Strip mining is sometimes used when the material is near the surface. The unwanted rocks are scraped or blasted away. Many materials are deeper underground. They are reached by digging deep shafts and tunnels.

▽ Strip mines are often used when the material is near the surface. They may ruin a beautiful landscape.

△ Gold is sometimes found mixed with sand in a riverbed. This person is using a pan to try to find some gold.

Atlas Copco

Using Earth

People have made things with earth for many centuries. Clay is the best material in soil for making things. Clay is made up of very small particles of rock and soaks up water easily. Wet clay is easy to shape. It can be made into many things, such as bricks for building and pots for storage. The finished items are dried in the sun or fired in a **kiln**.

▽ Pots harden in the heat of a kiln. Their shape is then permanent.

▷ In many parts of the world, house walls are made of mud.

◁ The most modern houses still use bricks and tiles made from clay.

▷ Sculptors often use clay because it can be shaped and molded so easily.

△ Porcelain is a very fine kind of pottery, made mainly from white china clay.

Breaking Up the Rocks

The Earth's rocks are constantly being broken up and worn away. The wind picks up particles of dust and sand and hurls them against the rocks. The rain beats down. A river can cut a valley into the hardest rock. Glaciers, rivers of ice in the mountains, can do the same. Slowly huge rocks are changed into the tiny particles that make up the soil.

▷ This rock has been worn into this shape by the wind and sand.

▽ Glaciers are rivers of ice. They gouge deep valleys out of the rock as they slowly move down the mountain.

▽ Changes in temperature from very hot to very cold can make rocks split and crack.

▽ Lichens grow on rocks. They, too, help to break up the rock's surface.

Soil

Soil is made when rocks are broken down into tiny particles. There are different types of soil. Each has rock particles of a different size. From the largest to the smallest they are gravel, coarse sand, fine sand, silt, and clay. Between the particles are air spaces. These can fill with water. When there is water, plants can take root and grow.

▽ If you dig down into the earth, you will see different layers in the soil.

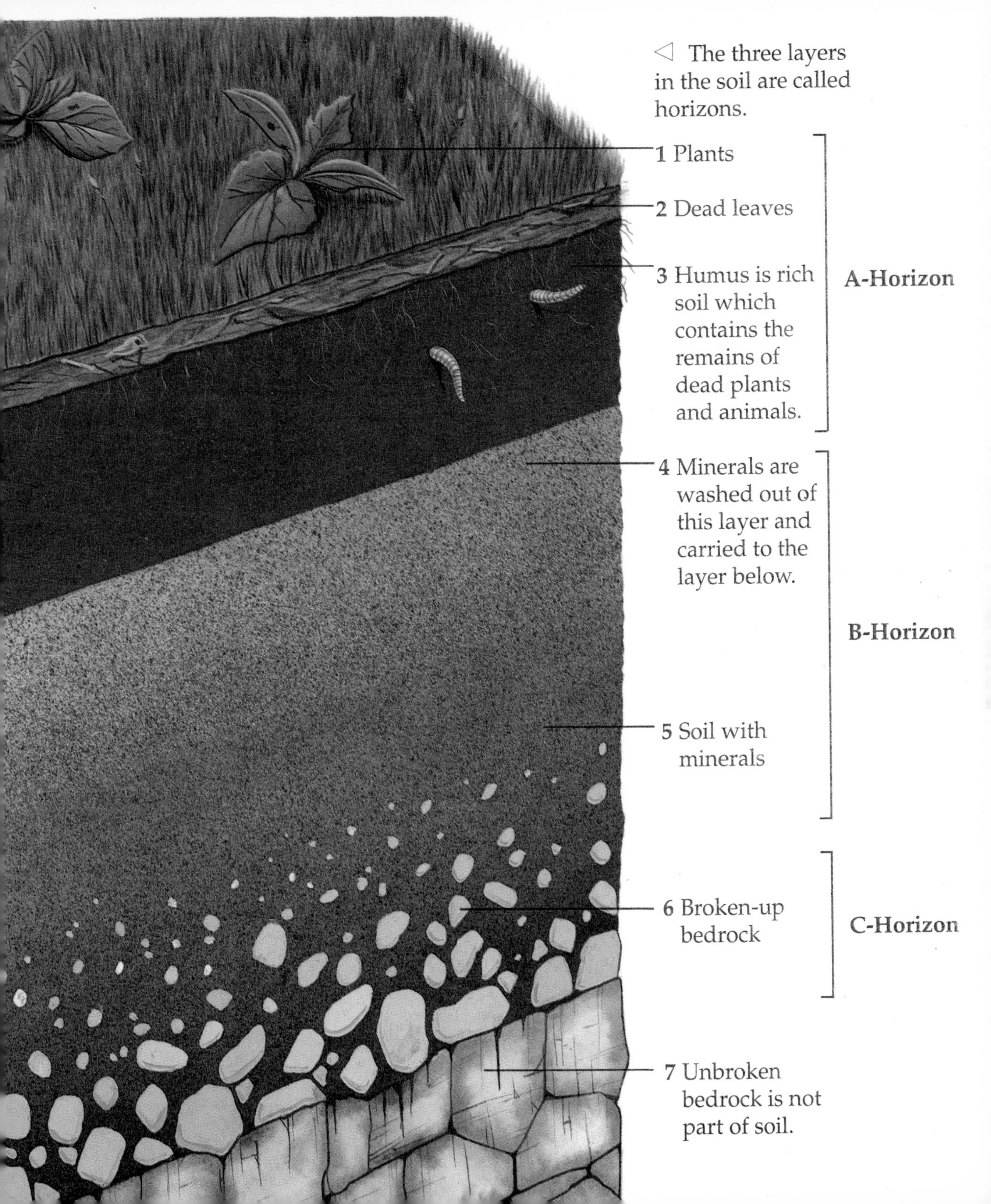

◁ The three layers in the soil are called horizons.

Soil and Plants

There are thousands of different kinds of plants in the world. What grows where depends largely on the kind of soil. Some plants can grow in dry, sandy soil; others in fine, wet soil. The roots of plants help to bind the soil together. They prevent it from being washed or blown away. When plants rot, their **nutrients** enter the soil and feed other plants.

▽ Rain forest soil is not very thick. So, some tall trees have buttress roots to support them.

▽ The shallow roots of some desert plants collect little rainwater from over a large area.

Buttress roots

▷ Reeds and grasses grow well in the water-logged soils of a marsh.

▽ The rich, thick soil in some parts of the world can support trees with deep roots.

Soils of the World

There are several different types of soil in the world. The kind of soil a place has depends on the type of rock that lies beneath it. It also depends on the kind of weather the area has had in the past and the kind it has now. The kind of plants that grow there and the shape of the landscape will also make a difference.

▷ Rain has washed the nutrients out of the soil. The land can only be used to graze sheep.

▷ In coniferous forests the pine needles break down very slowly on the forest floor. The soil is usually poor in nutrients.

▽ In some deserts, the dry soil is easily blown away by the wind. This leaves just rocks.

▷ On flat, fertile land farmers can grow important food crops like wheat.

Life in the Soil

A handful of soil may look lifeless, but it is not. A small amount of soil contains millions of tiny things called **bacteria**. It is also home to animals, such as mites, millipedes, and earthworms. These are nature's **recyclers**. They help to break down dead plant and animal remains. Earthworms help to mix the soil as they burrow through the ground.

▽ Without earthworms the soil becomes hard and lifeless. They tunnel through, letting air into the soil, and help water drain away.

◁ Larger animals, such as prairie dogs, make their homes in the soil.

△ Here are a few of the many animals that live in or on the soil.

Farming the Soil

▷ Teams of combine harvesters are used to harvest wheat in some parts of the United States.

Wherever there is soil, people grow things. Many different kinds of crops are grown for food. Before they are planted, the ground is broken up. This allows water and air into the soil. Large areas of land are plowed. Plants need water to grow. In dry areas, water can be brought to the fields by ditches or pipes. This is called **irrigation**.

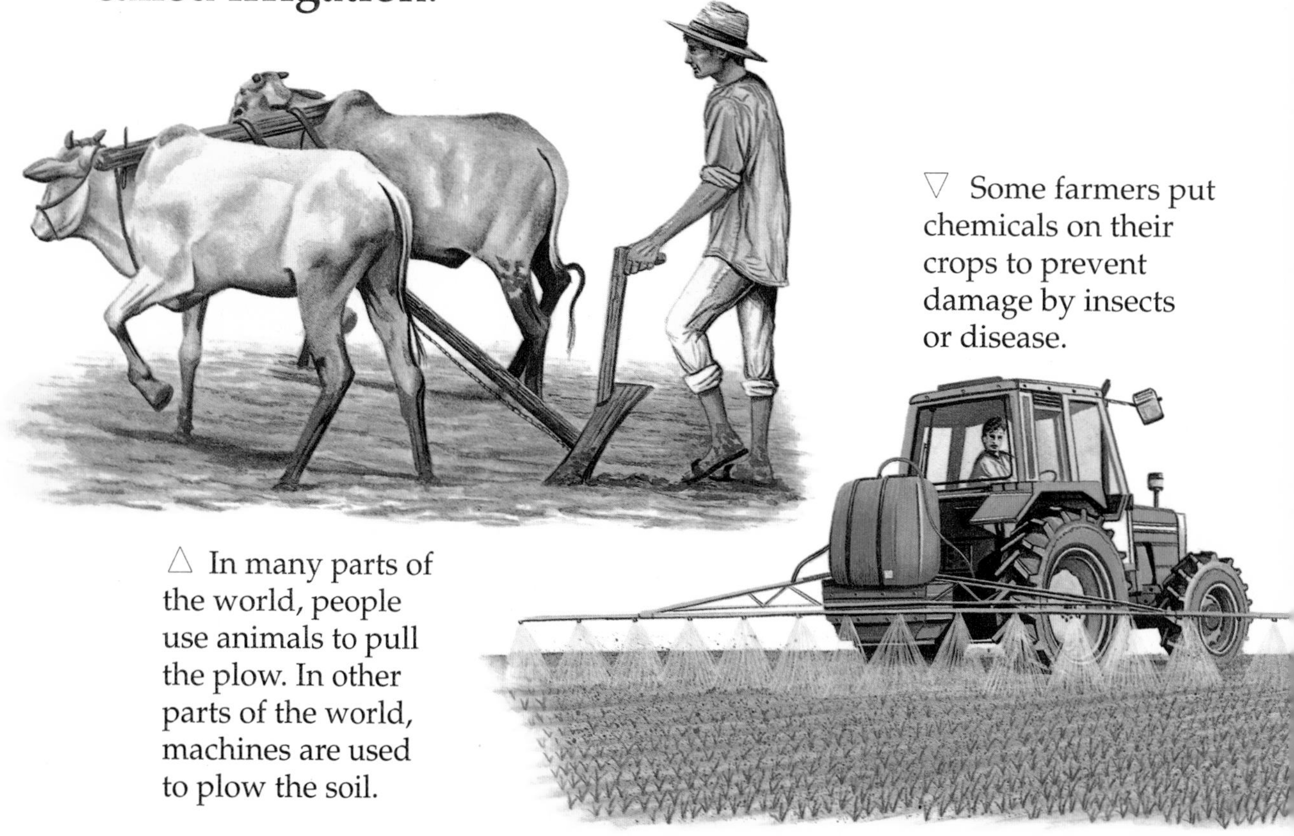

▽ Some farmers put chemicals on their crops to prevent damage by insects or disease.

△ In many parts of the world, people use animals to pull the plow. In other parts of the world, machines are used to plow the soil.

MF

Taking Care of the Soil

An area without plants can lose its soil. The wind and the rain will blow and wash it away. This is called **erosion**. Some chemicals we put onto our crops can make the soil and the rivers nearby poisonous. This is a form of **pollution**. Many people are now trying to grow crops without chemicals. Others hope to stop erosion by planting new forests.

▷ These farmers are making ridges along their fields to stop the rain from washing the soil away.

◁ These people are planting new forests to stop the soil from being washed away.

▷ After a heavy rain, a river can be carrying so much soil that it turns brown.

▽ Organic farmers use only natural fertilizers on the land.

Things to Do

- Plant your own garden. You can make an indoor garden by first putting a layer of gravel in a tray, then a layer of moist soil. Grass seeds, cacti, small bulbs, and flower seeds will grow if you give them enough light and water.

- Make a collection of different kinds of rocks or fossils or anything else that comes from the earth. You can try identifying them with a field guide book and then make a display of them.

- Write to your state's soil conservation department or your county cooperative extension agent. Ask for information on the kinds of soil in your region.

Glossary

bacteria Very tiny living things which can only be seen with a microscope.

chemicals Everything is made of chemicals. Some chemicals can be dangerous if there is too much of them and they are in the wrong place.

crops Plants we grow for food.

erosion A gradual wearing away of rock or soil by wind, rain, or ocean.

fuel A substance that releases energy, usually when it is burned. Coal is a fuel.

irrigation Taking water to a dry place in order to help crops grow.

kiln An oven in which bricks or pottery can be baked until hard.

lava Molten rock which, when it cools, becomes solid igneous rock.

minerals The substances that make up rocks. Some contain metals.

nutrients The substances which living things need to keep them alive.

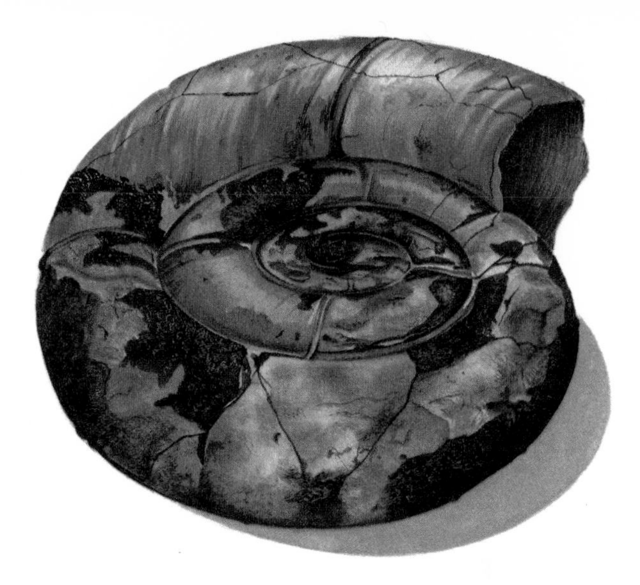

plates Any of the large movable pieces of rock that form the Earth's surface.

pollution Wastes and other harmful substances that spoil the air, soil, or water.

pressure The force of one thing pushing against another.

recycler A creature that helps to break down waste so that it can be used again.

Index

Photographic credits: John Cancalosi/Bruce Coleman Limited 8; Chris Fairclough Colour Library 14; Keith Gunnar/Bruce Coleman Limited 17; Robert Harding Picture Library 3; P. Morris 11; Stephen Oliver 18, 24; Prato/Bruce Coleman Limited 29; Zefa Picture Library 5, 7, 21, 23, 27.